Save £1000's in 15 minutes or less

Introduction

Introduction

I speak to many businesses every year that want to save money on their gas and electricity but discover that they can't because they have been tied into an expensive contract without knowing about it. I know that electricity and gas contracts may be boring and may not be your biggest worry, but by taking 15 minutes to read this guide you could save your business thousands of pounds.

As an independent broker I would like to help your business reduce it's energy costs, but even if don't use me as your broker, having read this guide that at least you will be in a position to make sure that you have the choice on who supplies your electricity and what you pay for it.

If you do nothing except make sure you cancel your current supply agreement in time, you will be in a position to shop around for the best price and you will save money.

Existing customers pay more than new customers of the same supplier.

To find out more about The Solutions Group and how we can help you, visit our website at www.energybrokers.co.uk, or contact me, nick@energybrokers.co.uk and I'll call you right back.

Loyal Customers Pay MORE

1. Loyal customers pay more - to get the best deal businesses need to switch suppliers

In the UK, 6 major electricity companies supply more than 96% of all businesses. When it comes to existing supply agreements, business customers are getting a raw deal from their electricity suppliers. Instead of being rewarded for staying with their supplier, they pay up to 100% more than new customers of the same supplier. Gone are the days when loyal customers were valued and given extra bonuses for their loyalty.

Business electricity suppliers have adopted this worst practice - 'new customer only' pricing polices.

2. Switching Electricity Supplier

Changing your electricity supplier should be quick and simple. Over 150,000 electricity supplies are changed each month. There is never an interruption to supply - all suppliers use the same wires and cables, the emergency services remain exactly the same.

But in order to change supply it is important that businesses understand their contractual obligation with their current supplier to ensure that they do not get trapped in expensive supply agreements.

3. Why Switch Supplier

Electricity is usually provided to businesses under two types of supply agreement:

- 28/30 Day Electricity Supply Contract
- Fixed Term Electricity Supply Contract

Your current supplier will only provide you with their Existing Customer price which frequently is massively inflated in relation to their New Customer prices.

Quite simply you need to switch to get the best deal and save your business money.

4. Your contractual obligations under a 28 Day Electricity Supply Contract

About 500,000 businesses are supplied under this form of contract and are usually businesses that have not switched supplier since the electricity market de-regulated.

The bad news is that it is estimated that customers on these supply agreements are being charged up to 100% more than new customers from the same supplier. Prices for these customers have been seen as high as 14 - 21 pence a unit of electricity compared to around 7.8 pence a unit for prices offered to new customers. (prices correct – Oct 2007)

By simply switching supplier, businesses can reduce their electricity bill by up to 50%.

The good news is that businesses with this form of supply agreement can switch supply quickly by providing their current supplier usually by just giving 28 days written cancellation notice.

See Appendix A and B for an example Cancellation Notice and contact details of suppliers.

The disadvantage of this form of supply agreement is that the supplier may increase your electricity prices at short notice usually of only 28 days. It is unusual for price decreases to be offered!

5. Your contractual obligation under a Fixed Term Electricity Supply Contract

The advantage of a Fixed Term Electricity Supply Contract is that usually the price is fixed during the agreed term of the contract. Fixed term contracts are available for duration of between 1 and 5 years. How long you should fix your prices for is really dependent on the current wholesale market condition, which is the key driver for the retail price of electricity. For example during 2005 & 2006 wholesale electricity was subject to unprecedented price rises, doubling over the period. This was not a good time to enter into a long-term contract.

After the wholesale electricity market peak of 2006 the price has crashed by about 60%, this made longer term supply agreements a good way of protecting your business from further price increases. With prices increasing again in 2008, shorter term contracts are becoming increasingly attractive.

Unfortunately your obligations under a Fixed Term Electricity Supply Agreement will vary from supplier to supplier. Later in this guide a summary of these obligations by supplier has been provided.

However, there are basic principles that are consistent across all suppliers:

Most businesses are on Evergreen Contracts that renew automatically unless broken by the customer. In other words, businesses need to give notice in advance to avoid automatic renewal, and a potential price hike.

In simple terms this means if you sign up for a 1 or say 3-year supply contract this agreement never ends unless you cancel in line with the supplier's specific cancellation notice requirements. A summary of the different supplier's cancellation requirements is provided in table 1 below.

Table 1: Suppliers cancellation requirements.

British Gas (British Gas Business/Scottish Gas/ Electricity Direct/Enron/Scottish Gas Business)	Written notice at least 90 days prior to the end of the initial contract.
EDF Energy (SWEB Energy/Seeboard Energy/ London Energy)	Written notice at least 28 days prior to the end of the initial contract.
Npower (Midlands Electricity/Yorkshire Electricity/ Northern Electric)	Written notice at least 90 days prior to the end of the initial contract.
Powergen (Economy Power/Norweb, Eastern Electricity/East Midlands Electricity/ Independent Energy)	Notification of cancellation can be given during the "Review Period". The Review Period is defined by Powergen as: "The Review Period is a period of not less than 14 days from the date of our written notification. Written notification will be not less than 30 days before the end of the Fixed Price Period" Notification can be given by: • Letter • Fax • E-mail
Scottish & Southern Energy (Swalec/Southern Electric/Scottish Hydro Electric/Atlantic Gas & Electricity)	Written notice not less than one calendar month before the end of the initial contract.
Scottish Power (SP Manweb/South of Scotland Electricity)	Written notice at least 60 days prior to the end of the Initial contract.

6. When is a Fixed Term Contract not a fixed price?

You would expect that having agreed to a fixed term electricity supply agreement the price was fixed for the duration - well think again. With certain suppliers their terms and conditions provide them with the option to increase prices.

Table 2: Example of where fixed term is not fixed price.

Npower (Midlands Electricity/Yorkshire Electricity/ Northern Electric)	"We reserve the right to increase the Contract Price on the 1st April of each Contract Year or on each anniversary at our discretion, to be advised by us from time to time, by the percentage rise in the Retail Price Index"

7. Your contractual obligation under a Fixed Term Electricity Supply Contract regarding price increase and renewal

As stated above, most Fixed Term Electricity Supply Agreements are EVERGREEN that means that if you do not serve your supplier with a valid cancellation notice as defined in table 1 above, you have entered a contract for life.

Price increase letters are issued by suppliers giving a variety of contractual obligations, which are listed in table 3 below.

These letters are cunningly disguised not to look like a price increase notification and are not headed Price Increase Notification but are instead headed as follows:

British Gas	"We'll protect your new prices for two years"
Npower	"Secure your electricity prices for 12 months and avoid future increases"
Scottish & Southern Electricity	"Electricity renewal offer"
EDF Energy	"Offering your business more"
Powergen	"Your electricity renewal offer"

Wolf in sheep's clothing!

These price increase letters are so successful that a leading supplier retains 95% of its customers.

Why?

Most businesses do not recognise the ramification of these letters from their suppliers. The letters do not say what the % increase is, how much more it will cost you, or what the increase will be to your direct debit.

They refer to unit prices. Now we are sure that you do not keep what you pay per unit of electricity under your pillow, so when a letter drops on your desk saying "We'll protect you, we've taken steps to protect you… your new guaranteed prices are printed overleaf". Overleaf is just a unit rate of 14 pence - what does this mean to you?

The Daily Telegraph investigated these practices and reported:

"British Gas Accused Over Tactics on Renewal

British Gas is being accused of using "crafty tactics to pressure small businesses into signing up early for new energy contracts at substantially higher prices......... Some companies on year-long contracts face rises of 20pc-30pc after big increases last year but businesses coming to the end of a longer-term deal are confronted, in some cases, with a near doubling in price... this is pretty crafty tactics to put up prices. In many cases customers simply accept the price hike through inertia because they're too busy trying to run their own business"

Although the format of the price increase letter varies from supplier to supplier they all do have some common themes:

• You do not have to give acceptance - the letters say that the offer will have been deemed to have been accepted unless the customer rejects the price increase.
• You do not even have to receive the letter - your supplier's obligation is fulfilled simply by sending the letter.
• Some extend the term by 24 months.

Putting it simply if you receive one of these letters and do nothing you will have entered into a legally binding contract for a further supply period at the increased rates.

So if you thought that you had entered into an agreement with your electricity supplier that at the end of the agreed supply period your contractual obligation would end and you would be free to shop around, you are mistaken.

Table 3 below outlines each of the big 6 supplier price increase practices:

British Gas (British Gas Business/Scottish Gas/Electricity Direct/Enron/Scottish Gas Business)	1) Price increase letter is sent 145 days before your initial contract end date. 2) From the date of the price increase letter your have 42 days to reject the price increase. 3) The letter includes updated terms & conditions. 4) Increased prices are a contractual obligation for a period of 2 years. 5) Do nothing and you will be deemed to have accepted the increased prices, a 2 year supply period and the new terms & conditions.
EDF Energy (SWEB Energy/Seeboard Energy/London Energy)	1) Price increase letter is sent 30 days before your initial contract end date. 2) The letter does not give you the option to reject the price increase. 3) If you have not given termination notice at least 28 days before the end of your initial contract period you do not have a legal entitlement to reject the price increase. 4) Increased prices are a contractual obligation for a period of 23 months.
Npower (Midlands Electricity/Yorkshire Electricity/Northern Electric)	1) Price increase letter is sent 33 days before your initial contract end date. 2) The letter does not give you the option to reject the price increase. 3) If you have not given termination notice at least 28 days before the end of your initial contract period you do not have a legal entitlement to reject the price increase. 4) Increased prices are a contractual obligation for a period of 1 year. 5) Do nothing and you will be deemed to have accepted.
Powergen (Economy Power/Norweb, Eastern Electricity/ East Midlands Electricity/Independent Energy	1) Price increase letter is sent 41 days before your initial contract end date. 2) From the date of the price increase letter your have 14 days to reject the price increase. 3) The letter includes updated terms & conditions. 4) Increased prices are a contractual

	obligation for a period of 1 year. 5) Do nothing and you will be deemed to have accepted the increased prices, a 1 year supply period and the new terms and conditions.
Scottish & Southern Energy (Swalec/Southern Electric/Scottish Hydro Electric/ Atlantic Gas & Electricity)	1) Price increase letter is sent 65 days before your initial contract end date. 2) From the date of the price increase letter you have 17 to 38 days to reject the price increase. 3) The letter includes updated terms & conditions. 4) Increased prices are a contractual obligation for a period of 2 years. 5) Do nothing and you will be deemed to have accepted the increased prices, a 2 year supply period and the new terms and conditions.
Scottish Power (SP Manweb/South of Scotland Electricity)	1) Price increase letter is sent 38 days before your initial contract end date. 2) From the date of the price increase letter your have 7 days to reject the price increase. 3) Increased prices are a contractual obligation for a period of 1 year. 5) Do nothing and you will be deemed to have accepted the increased prices for a 1 year supply period.

What is your MPAN

1. What is your MPAN?

An MPAN (Meter Point Administration Number) is a unique number to the property. It is found on the electricity bill issued issue by the supplier. This is sometimes called a Supply Number but it should not be confused with the customer reference number. It is not displayed on the actual meter.

The full MPAN is 21 digits in length and should be printed in the format below on the electricity bill for the supply.

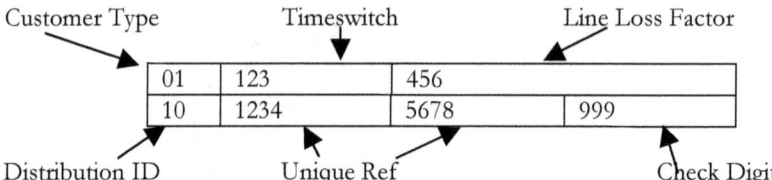

Customer Type	Timeswitch	Line Loss Factor	
01	123	456	
10	1234	5678	999

Distribution ID   Unique Ref   Check Digit

Suppliers can also obtain the MPAN on the behalf of customers by using the MPAS

Customers wishing to find out their MPAN you can also call their local Distribution Company and they will be able to provide the 13 digit MPAN core over the phone. Full details can be found in Appendix D.

In most cases this should be enough information to arrange the transfer of an electricity supply, it is also worthwhile asking that they send the information by post - that way you will also receive the full MPAN.

2. What does the MPAN mean?

a) Profile Class

Every property has a profile class.

Profile classes are used where half-hourly metering is not installed and provides the electricity supplier with an expectation as to how electricity will be consumed throughout the day.

01 Domestic Unrestricted
02 Domestic Economy 7
03 Non-Domestic Unrestricted
04 Non-Domestic Economy 7
05 Non-Domestic Maximum Demand 0-20% Load Factor
06 Non-Domestic Maximum Demand 20-30% Load Factor
07 Non-Domestic Maximum Demand 30-40% Load Factor
08 Non-Domestic Maximum Demand >40% Load Factor

Your profile class will give suppliers an idea of your consumption patterns and your efficiency at using the energy you consume. This in turn is reflected in the standing charge and unit rates that you will be charged. As a general rule, higher usage needs a higher profile.

So shops, small offices, etc are fine on an 03 / 04 tariff, whereas larger users are better off on higher numbers, 05 to 08, and the higher the number the better (lower unit costs).

If there has been a significant change is usage, with may be worth asking your supplier to review your profile.

b) Meter Time Switch Code (MTC)

The Meter Time Switch Code indicates how many registers (set of meter reads or dials) the electricity meter has and what times they will operate during the day. The Meter Time Switch Code will show if the meter has two registers, one that records day consumption, the other night. Numbers 501 – 799 indicate that there are related meters.

c) Line Loss Factor (LLF)

The Line Loss Factor code stipulates the expected costs the distribution company will charge the supplier for using the cables and network in the region. This Line Loss Factor code will also indicate to the electricity supplier the potential charges incurred, due to loss of energy incurred whilst getting the electricity supply to the meter.

d) Distributor ID

The Distributor ID will identify the local Distribution Company for the electricity supply. The Distribution Company is responsible for management of the distribution system and electricity wires, which transports the electricity to the meter.

10 - Eastern Electricity
11 - East Midlands Electricity
12 - London Electricity
13 - MANWEB
14 - Midlands Electricity
15 - Northern Electricity
16 - NORWEB
17 - Scottish Hydro-Electric
18 - Scottish Power
19 - Seeboard
20 - Southern Electricity
21 - SWALEC
22 - SWEB
23 - Yorkshire Electricity

e) Meter Point ID Number

This is a unique number within the distribution area to identify the actual meter.

f) Check Digit

This number is calculated from the Distributor ID and Meter Point ID Number to provide a check digit that other systems can use to validate the MPAN.

Business Gas

1. Do it now

This is the most important piece of advice we give our clients:

Ensure that you have sent your termination notice into your existing supplier. Don't wait for the renewal offer, don't wait until 90 days before the contract ends. Just do it, whether you've had your renewal notice or not. Suppliers have been known to send out renewal offers after the notice period has expired – meaning that waiting for the renewal can tie you into a contract whether you like it or not.

2. Your Contract Termination

It is important to terminate your current gas contract with your present supplier as failure to terminate within the suppliers notice period can result in the contract being rolled over for a further 12-month period. Most gas suppliers have a 90-day notice period as part of their terms and conditions, but we recommend that you check with your supplier by ringing the customer service number that appears on your bill. By terminating the contract with your current supplier, doesn't mean that you can't accept a renewal offer from them, or that you'll be cut off when the contract comes to an end but it allows you the freedom to search and compare the market. We have provided an example of the gas termination letter in appendix E, which will need completing on your company letterhead, before sending by recorded delivery.

3. Gas Prices

Unlike the electricity market your gas is pre-bought and sold on by the energy companies. Your price will depend upon two main factors: your usage, which the companies are given by the network operator (Transco), and delivery charges. The further away from the terminals you are, the higher transport costs you incur.

If your usage is in below ~750,000 kWh per year the prices will come from the gas companies general pricing structures which used to move every week, but now change almost daily. If your usage is above this level and into the 1,000,000's the companies will go to the open market and get you a price that may be valid for only a day or few hours.

The day to day prices in the larger market reflect the open market conditions, if the Met Office tell us 'its going to be the worst winter in history' (see Nov 2005), if there is trouble in the middle east, if Russia cuts off a neighbour, the prices will fluctuate dramatically. Often there is not a 'real' supply issue; it is just the nerves of the market.

4. MPRN (M number, Meter Point Reference Number)

Every mains gas meter in the UK has an MPRN. The MPRN is unique to the meter and therefore the property and does not change if you change supplier or even if you move.

Sometimes the MPRN is referred to as a M Number or Meter Point Reference Number

Your MPRN should be printed on a recent gas bill. Most suppliers print the MPRN on either the first page of their gas bills or the reverse of the front page.

Your MPRN should not be confused with your gas supplier account number or the meter serial number printed on the meter itself. Some very recent gas meter installations have the MPRN clearly identified on the meter itself although this practice is quite rare.

The format of a MPRN is quite standard and consists of between 6 and 10 numerical digits only.

If the MPRN starts with 74 or 75 then you should be aware that an Independent Gas Transporter pipes gas to your property. This may have an adverse effect on the price you pay for gas as some gas suppliers charge a premium rate to customers served by an Independent Gas Transporter.

If your MPRN is not printed on a recent gas bill or if have just moved into your home and do not yet have a gas bill, you can find out your MPRN by calling Tranco's MPRN helpline on 0870 608 1524 (Transco can only provide details of MPRNs that do not belong to Independent Gas Transporters – contact details of which can be found in Appendix F.)

5. Independent Gas Transporters – hidden costs

If you are served by an Independent Gas Transporter's (IGT) network (i.e. not Transco) you may find that the price you pay for gas is slightly higher than the standard published price provided by the gas supplier.

This is because the gas supplier must pay both Transco and the Independent Gas Transporter to deliver gas through the pipes to your business.

In practical terms there is nothing wrong with having an Independent Gas Transporter deliver the gas to your business, the downside is the additional cost element.

Many construction companies use Independent Gas Transporters to save money when they are building new properties, as the Independent Gas Transporters will offer, in most cases, a more competitive price for all the connection work than Transco.

The problem occurs when you then wish to compare the prices offered by the gas suppliers. Some of the low cost gas suppliers will not quote for a supply to property connected via an Independent Gas Transporter and the majority of other suppliers will charge a premium.

So whilst the builder gets a good deal when installing the gas network, you the consumer end up paying for this in increased costs forever.

To find out whether you are connected to the gas distribution system via an Independent Gas Transporter is quite simple.

If the M number (MPRN) shown on your gas bill is 10 digits long and begins with 74 or 75 then you are being supplied by an Independent Gas Transporter.

You should find your M number (MPRN) on either the front page or reverse of the front page of a recent gas bill.

Home Energy

1. Finding the best price

With energy companies now offering a bewildering range of different deals, it's easy to become confused about what's on offer. To help you decide which is best for you, use one of the price comparison services that are accredited by energywatch.

You can use any one of them to get a quote to find out how much you could save by switching to a new supplier.

Website	Phone
www.saveonyourbills.co.uk	0845 123 5278
www.homeadvisoryservice.com	0845 1800 300
www.unravelit.com	
www.moneysupermarket.com	0845 345 5708
energy.moneyexpert.com	
www.ukpower.co.uk	0845 009 1780
www.switchwithwhich.co.uk	0800 533 031
www.theenergyshop.com	0845 330 7247
www.uswitch.com	0800 404 7908
www.energylinx.co.uk	0845 225 2840
www.confused.com	
www.energyhelpline.com	0800 074 0745

These independent price comparison sites help thousands of households every month, to compare prices and switch to better deals. They offer a comprehensive service that compares your current supplier, tariff and usage with the tariffs from all the major suppliers, to help find the best deal for you and provide you with a free and easy to use switching service. They also give detailed information on each tariff, including gas and electricity unit prices and any discounts.

2. How do I change my supplier?

Once you are happy that you have selected the supplier best suited to meet your needs, changing is a fairly simple process:

1. Get in touch with the new supplier and agree a contract with them. Once the contract is agreed, the transfer process should take about six weeks to complete. Your new supplier will keep you informed about how your transfer is being progressed.
2. Pay any outstanding bills owing to your existing supplier. If you do not, they may prevent you from transferring.
3. Take a meter reading on the day you change supplier. If your old supplier does not use it to work out your final bill, or your new supplier does not use it as the starting point for your first bill, let them know the meter reading you have taken.

3. Can't decide which company to switch to...?

Then you might also be interested to see how the suppliers compare with each other - take a look at the statistics that show the number of cases energywatch has received from each supplier's customers.

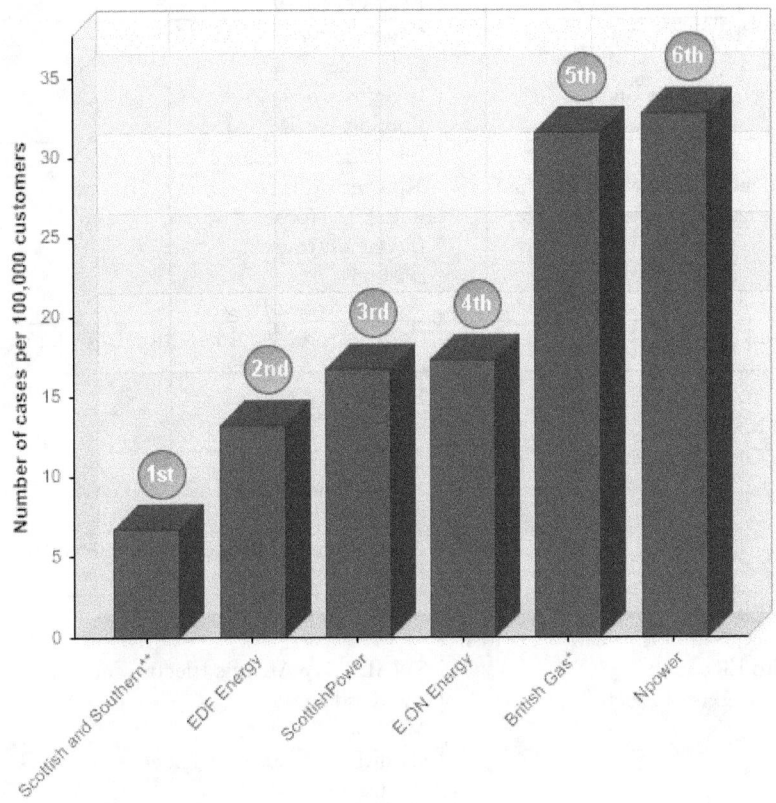

1. Average domestic cases to energywatch per 100,000 customers by supplier (December 2007 - February 2008)

2. 1st indicates best performing company, 6th indicates worst performing company

3. * includes Scottish Gas

4. ** includes Atlantic, Scottish Hydro Electric, Southern Electric, SWALEC

Appendix A - Supplier contact details for cancellation

British Gas (British Gas Business/Scottish Gas/Electricity Direct/ Enron/Scottish Gas Business)	Cancellation Method: Letter Spinneyside Penman Way Grove Park Leicester LE19 1SZ
EDF Energy (SWEB Energy/Seeboard Energy/London Energy)	Cancellation Method: Letter SME Sales Team Freepost 3814 London WC1V 6AJ
Npower (Midlands Electricity/Yorkshire Electricity/ Northern Electric)	Cancellation Method: Letter Npower Business Radcliffe House Blenheim Court Solihull West Midlands B91 2AA
Powergen (Economy Power/Norweb, Eastern Electricity/ East Midlands Electricity/Independent Energy)	Cancellation Method: Letter, E-mail, Fax Contract Termination PO Box 9042 Sherwood Park Annesley Nottingham NG15 5AZ E-mail: smecontractterminations@powergen.co.uk Fax: 0115 877 5755
Scottish & Southern Energy (Swalec/Southern Electric/Scottish Hydro Electric/ Atlantic Gas & Electricity)	Cancellation Method: Letter SWALEC & Atlantic Electric and Gas Ty Meridian Malthouse Avenue Cardiff Gate Business Park Pontprennau Cardiff CF23 8AU Scottish & Southern Energy Centenary House 10 Winchester Road Basingstoke Hants RG21 8UQ
Scottish Power (SP Manweb/South of Scotland Electricity)	Cancellation Method: Letter SME Contracts Manger Section 12 Cathcart Business Park Glasgow G44 4BE

Appendix B - Contract cancellation template letter

To:

From:

Date:

A/C No.:

Dear Sir,

MPAN Number(s):

S					S				

S					S				

S					S				

We have decided to switch electricity supplier.

Our supply from you will end on ___ / ___ / ____. Please accept this letter as our formal cancellation notice in accordance with your terms and conditions.

Yours faithfully,

Appendix C - Summary of supplier's terms & conditions of supply

British Gas (British Gas Business/ Scottish Gas/Electricity Direct/ Enron/Scottish Gas Business)	1.7 (A) At the end of the Initial Period this agreement shall remain in full force and effect for successive periods of one year or such other period as notified by the Supplier and agreed by the Customer ("Renewal Periods"), each of which shall begin either on the date of the Initial Period ends or on the relevant anniversary of that date. (B) Either the Customer or the Supplier may terminate this agreement at the end of the Initial Period or Renewal Period by giving to the other written notice at least 90 days prior to the end of the Initial Period or Renewal Period, as the case may be.
EDF Energy (SWEB Energy/Seeboard Energy/London Energy)	3.2 We may change the terms and conditions of your supply contract (including the charge agreed within your supply contract) from time to time. Unless otherwise permitted by these terms, we shall not increase the prices we have agreed with you prior to the first renewal date. 6.1 You can end this contract by giving us written notice to end this contract with effect on the first renewal date (or if this contract has continued beyond that date, with effect from the next anniversary date), such notice to be received by us 28 days in advance of the day it is to take effect and must be sent to: London Energy Customers SME Sales Team, Freepost 3814, London WC1V 6AJ Seeboard Energy Customers SME Sales Team, Freepost 3815, Hove BN3 5AW
Npower (Midlands Electricity/ Yorkshire Electricity/ Northern Electric	2.6 Subject to earlier termination as hereinafter provided this Contract shall initially remain in force for the fixed period as defined on the front sheet of this Contract as the "Initial Term" which shall commence on the Effective Date. Following the termination of the Initial Term this Contract may be brought to an end by either party giving the other not less than 90 days notice in writing prior to the end of the Initial Term or the end of any subsequent anniversary of the end of the Initial Term (as the case requires). If no notice is given the Contract will continue for successive periods of 24 months. The Contract price during such periods may at our discretion be changed to those for new contracts let by us for premises similar to the premises supplied under this Contract applicable at the end of the Initial Term or at the relevant anniversary of the end of the Initial Term (as the case requires). The Contract shall

	otherwise continue on the same terms and conditions. 4.4 We reserve the right to increase the Contract Price on the 1st April of each Contract Year or on each anniversary of the Effective Date at our discretion, to be advised by us from time to time, by the percentage rise in the Retail Price Index (all items) over the 12 months prior to the end of the month immediately preceding the relevant anniversary of the Effective Date or the 1st April in the relevant Contract Year as applicable (the "RPI Contract Price Rise") provided that we shall not invoke our right to a RPI Contract Price Rise on the 1st April of the first Contract Year. Should the Retail Price Index cease to exist, we shall substitute another index of similar nature.
Powergen (Economy Power/Norweb, Eastern Electricity/ East Midlands Electricity/ Independent Energy)	4. If you are in a fixed price contract you can only end this contract during the review period. When ending the contract, you must notify us by one of the following methods: in writing to Contract Terminations, PO Box 9042, Sherwood Park, Annesley, Nottingham NG15 5AZ by e-mail to smecontractterminations@powergen.co.uk by fax on 0115 877 5755 6. The Review Period is the opportunity for you and us to agree the price and terms that will apply at the end of the current Fixed price Period or Fixed Term Contract: You do not need to provide any form of termination prior to the Review Period, as you will be free to discuss your needs during this period. If you do provide a termination notice outside of the Review Period, we will not take account of it during the Review Period. We will write to you not less than 30 days before the end of the Fixed Price Period or Fixed Term Contract with our offer of terms for the continuation of Services. The Review Period is a period of not less than 14 days from the date of our written notification, during which you have the opportunity to discuss this offer with us. If you wish to discuss this offer, you need to telephone us within the Review Period on the number provided in the letter we send you. You can only end a Fixed Term Contract during the Review Period at the end of its term
Scottish & Southern Energy (Swalec/Southern Electric/ Scottish Hydro Electric/	4.4 Subject to clauses 4.6, 4.7 and 7.1 upon giving you not less than 14 days notice in writing We shall be entitled to vary the Prices with effect from the First Termination or anniversary thereof. 4.5 In the event that the period between the Commencement Date and the First Termination Date is 36 months or more

Atlantic Gas & Electricity)	then We shall upon giving You not less than 28 days written notice be entitled to vary the prices in accordance with the retail price index calculated in clause 4.7 on the third anniversary from the Commencement Date of this Agreement and on every subsequent anniversary thereafter. 4.7 With effect from the First Termination date or anniversary thereof and only in the absence of any effective notice in accordance with clause 4.4 or 4.6 all Prices in the Schedule will be increased by the retail price index. 7.1 You can terminate this Agreement by giving Us not less than one calendar month's written notice to expire on the First Termination Date or any anniversary thereof.
Scottish Power (SP Manweb/South of Scotland Electricity)	5. 5 You may terminate this agreement (a) if there is an Earliest Termination Date on that Date or any anniversary of that Date by giving to us notice in writing to that effect at least 60 days prior to the proposed date of termination (b) if there is no Earliest Termination Date but there is a Review Date on the Review Date or any anniversary of the Review Date by giving to us notice in writing to that effect at least 60 days prior to the proposed date of termination.

Appendix D – MPAS Contact Details

The telephone numbers for each Distribution Company's MPAS service is noted below.

Region	Distribution Company	MPAS Number
Eastern	EDF Energy	0870 1963082
East Midlands	East Midlands Electricity Distribution	0870 6070459
London	EDF Energy	0845 6000102
North Wales, Merseyside and parts of Shropshire	SP Manweb	0845 2709101
West Midlands	Aquila Networks	0845 0707017
North East	Northern Electric Distribution	0845 6013268
North West	United Utilities Electricity	0870 7510093
North Scotland	Scottish Hydro-Electric Power Distribution	0870 9009690
South Scotland	SP Distribution	0845 2709101
South East	Seeboard Power Networks	0845 6015467
Southern	Southern Electric Power Distribution	0870 9050806
South Wales	Western Power Distribution	0845 6015972
South West	Western Power Distribution	0845 6015972
Yorkshire	Yorkshire Electricity Distribution	0845 3300889

Ask for their MPAS department. They should be able to either tell you the full 21 digit MPAN or arrange to send it to you.

Appendix E – Gas Contract Termination Template Letter

To:

From:

Date:

A/C No.:

Dear Sir,

MPRN Number(s):

1) _____
2) _____
3) _____
4) _____

We have decided to switch gas supplier.

Our supply from you will end on ___ / ___ / ____. Please accept this letter as our formal cancellation notice in accordance with your terms and conditions.

Yours faithfully,

Appendix F - Independent Gas Transporters

IGT Name and Address	First eight characters of your MPRN range		IGT Telephone Number
	From	To	
British Gas Connections 30 The Causeway Staines Middlesex TW18 3BY	74400001	74550000	Tel: 0845 600 6311 Fax: 01784 874497
ES Pipelines Ltd Old Pump House Elmer Works, Hawks Hill Leatherhead Surrey KT22 9DA	74900001	74930000	Tel: 01372 227560 Fax: 01372 377996
The Gas Transportation Company Ltd Woolpit Business Park Bury St Edmunds Suffolk IP30 9UQ	74250001 75300001 76000000	74400000 75500000 76399999	Tel: 01359 240363 Fax: 01359 241902
Independent Pipelines Ocean Park House East Tyndall Street Cardiff CF24 5GT	74000001 75000001 76400000 76543215	74250000 75300000 76543213 76753605	Tel: 02920 908550 Fax: 02920 314140
Mowlem Energy Ltd 2 Redwood Court Peel Park East Kilbride Glasgow G74 5PF	74980001	74995000	Tel: 01355 909600 Fax: 01355 909601
Quadrant Ltd Ocean Park House East Tyndall Street Cardiff CF24 5GT (Formerly East Midlands Pipelines Ltd.)	74850001	74900000	Tel: 02920 304040 Fax: 02920 314140
Scottish Power Gas Ltd	74650001	74750000	Tel: 01415674026

St Vincent Crescent Glasgow G3 8LB			Fax: 01415674274
SSE Pipelines Ltd 55 Vastern Road Reading Berkshire RG1 8BU	74750001 75510001	74800000 75550000	Tel: 01753 695631 Fax: 01189534695
United Utilities Network 12th Floor Oakland House Talbot Road Stretford Manchester M16 0HQ	74930001	74950000	
United Utilities Pipelines PO Box 3010 Links Business Park Fortram Road St Mellons Cardiff CF3 0DS	74950001	74980000	
Utility Grid Installations Mount Stuart House Mount Stuart Square Cardiff CF10 5FQ.	74995001 75500001	74999499 75510000	Tel: 02920 435320 Fax: 02920 435321

For any other MPRNs the transporter is Transco, whose address details are:

Transco plc
51 Homer Rd
Solihull
West Midlands
B91 3QL

Tel: 0121 626 4431   Fax: 0121 623 2625

www.ingramcontent.com/pod-product-compliance
Lightning Source LLC
Chambersburg PA
CBHW051421170526
45165CB00004BA/1911